R.A.A.F. BAIRNSDALE

THE STORY OF A WARTIME AIRFIELD

JIM PRENDERGAST

WITH A FOREWORD BY
AIR VICE-MARSHAL C.D. CANDY,
C.B., C.B.E.

First Published: March 1982.

Printed by James Yeates & Sons (Printing)
Pty. Ltd.
Bairnsdale. Victoria.

ISBN 0 9593363 0 3

FOREWORD

Air Vice-Marshal C. D. Candy, C.B., C.B.E., R.A.A.F. (Ret'd)

Historian, Hal Porter, in his book, "Bairnsdale — Portrait of an Australian Country Town" — (circa 1977) dealt in only the briefest terms with the role of the Royal Australian Air Force in East Gippsland during World War II.

This book, by Jim Prendergast, complements that by Hal Porter. It contains a wide and accurate coverage of the activities of the Royal Australian Air Force units and personnel based at Bairnsdale between 1939 and 1945.

One of the objectives of the Armed Services in democratic countries is to integrate, as closely as practicable, with the civilian communities amongst whom they are serving in peace and war. This tends to become one-sided in war for security reasons.

The author, through personal experience, has been able to demonstrate in this book how closely that objective went towards attainment in East Gippsland throughout World War II. When readers realise that during the period April, 1942 to March, 1943, one Royal Australian Air Force unit (No. 1 O.T.U.) expanded from a small beginning at Bairnsdale to the largest flying unit in the R.A.A.F., comprising some 100 operational aircraft and almost 3,000 personnel, they are well able to comprehend the enormous level of co-operation necessary from the local community. That this was so readily forthcoming on all occasions with understanding and without complaint is to the enduring credit of Bairnsdale. It is an enviable record of which East Gippsland can still be justifiably proud.

The introduction of the "30 years rule" has enabled Jim Prendergast to lift the veil of secrecy from the R.A.A.F. activities at the Bairnsdale aerodrome during World War II. This, he has done in an analytical and interesting style.

His book is being launched during 1982 — the 120th anniversary of Bairnsdale and the 40th anniversary of the establishment of No. 1 O.T.U. at Bairnsdale — which will ensure it receives the wide publicity it so richly deserves.

I commend it as exciting and refreshing reading to those who already know what took place behind the veil and to the many who have been waiting, with expectant anticipation, for the lifting of the veil.

C. D. CANDY, C.B., C.B.E.,
Air Vice-Marshal, R.A.A.F., Ret'd.,
former Commanding Officer,
No. 1 Operational Training Unit,
R.A.A.F. Bairnsdale and East Sale.

Beaufort A9-66 over R.A.A.F. Bairnsdale.
(From an original painting by the author)

PRE-WAR PLANNING

By the beginning of 1938 events in Europe made it very obvious that it was now only a matter of time before the outbreak of a second World War.

The Australian Government, in March 1938, announced a revised Three Year Defence Plan in which £12.5 million had been set aside to provide additional men and equipment for the R.A.A.F. First-line strength would be raised to 198 aircraft manned by 500 officers and 4,500 men.

Later in the year, following a report prepared for the Australian Government by Marshal of the Royal Air Force, Sir Edward Ellington, in which he was highly critical of the training methods and the obsolescent aircraft then in use by the R.A.A.F., the Air Force Budget was increased to about £16.5 million. From this, to overcome the grave lack of modern operational aircraft, fifty Lockheed Hudsons were ordered from the U.S.A. and the R.A.A.F. was also to receive a number of Avro Ansons from the British Government on a charter basis as an interim measure. The number of first-line aircraft planned, was raised to 212 planes.

Strategic thinking at that time was that even if Japan entered a war, Australia would be safe from bombing and that the major role of the R.A.A.F. in Australia would be to intercept any aircraft operating from enemy surface raiders. These were expected to be only small in numbers and of limited striking power. Equally important, however, would be the need to provide adequate protection for shipping passing along the Australian coast. To meet these contingencies, among several other items in the Budget, it was proposed that a chain of landing grounds be developed along the coastline for emergency use in time of war. These were to include landing fields at Bairnsdale and Mallacoota.

Negotiations began between the Premier of Victoria and the Commonwealth Prime Minister Joseph Lyons for the inauguration of defence and development work and the establishment of a number of country aerodromes in Victoria. Included in the financial structure was an amount of £17,000 allotted under the R.A.A.F. expanded Budget for the preparation of strategic landing grounds at Bairnsdale and Mallacoota. Funds were also made available at that time for the Civil Aviation Department to establish and develop aerodromes at Ballarat, Benalla, Apollo Bay and at Yanakie in South Gippsland.

An inspection of an area situated seven miles from the Bairnsdale Post Office by road to the south-west of the town on the Bengworden Road followed. On April 4, 1939, the Civil Aviation Department estimated that the preparation of a rectangular area, approximately 2,944 ft. by 3,000 ft. in the south-east corner of the area proposed to be acquired would cost £2,370. Although funds for its construction were being provided from Air Force funds, the aerodrome was to be under the control of Civil Aviation and the work on the aerodrome was to be carried out by that Department on a day labour basis.

By July, 1939, a tender for the preparation of an aerodrome at Mallacoota had been accepted at a cost of £5,294. Work commenced immediately.

On August 22, 1939, the Department of Interior invited tenders for the preparation of a landing ground at Bairnsdale, but executive action to commence construction was still not finalised when war was declared. Although several Government Departments had become involved, only a minor road running through the property had been closed and the Department of the Interior had been granted permission from the Bairnsdale Shire to remove trees from the road adjoining the proposed aerodrome. It became necessary to adopt extra-ordinary measures to meet the urgent requirement. Accordingly, verbal discussions with officers of the Civil Aviation Department resulted in immediate action and on Monday, September 4, 1939, just 24 hours after war commenced, arrangements were made at Bairnsdale for the employment of thirty men to commence clearing the landing ground and a larger number of men were employed later in the same week. This work was carried out under the direct control of the Department of Civil Aviation.

Now that a state of war existed, the landing ground at Bairnsdale became an urgent operational necessity, connected as it was with the protection of trade routes along the south-eastern coast. In November, 1939, the Acting Minister for Air, Mr Harold Holt, approved a recommendation from the Chief of Air Staff for the provision of thirty-eight Advanced Operational Bases (A.O.B.'s) along the Australian coast. These were to give extended range to aircraft engaged on coastal reconnaissance. They were to provide refuelling, re-arming and W.T. facilities, thereby increasing the radius of patrolling planes and ensuring a more adequate degree of protection to coastal shipping. Among the first ten to be classified as most urgent were those at Bairnsdale and Mallacoota and the necessary funds for their development were allocated.

Bairnsdale Shire Council was granted £3,000 for work on the gravelling of the runways in December, 1939, and in April, 1940, tenders were called for the erection of Wireless Hutments.

During 1940, the R.A.A.F. Development Programme called for a General Reconnaissance School to be established and Bairnsdale was selected as the suitable site as by then the existing airfield under control of the Department of Civil Aviation had runways extending 1,500 yards in four directions.

The proposed unit required a flying section complete with workshops, administrative and instructional buildings, equipment stores and hangars for 18 service aircraft. Accommodation would be needed for 58 officers, 43 sergeants and 263 airmen. The aircraft were to be housed in Bellman Hangars and the living accommodation was to be provided in standard temporary hutments with all engineering services as required. Additional land would have to be procured to accommodate the W/T buildings and to provide an area for sewage treatment.

The School was required to be in operation by July 1, 1940, and it was recommended to Air Board that the sum of £105,810 be allocated for the necessary buildings and services.

Plans were drawn up for the School and submitted to the Air Board on May 22, 1940. However, the Board decided that the question of the establishment of a General Reconnaissance School would have to be deferred in view of an anticipated shortage of aircraft.

Australia had entered the war with just 246 aircraft of which only 164 were rather optimistically regarded as operational. The most modern were 82 Avro Ansons and 7 Wirraways. The promised Hudsons had still not materialised from the United States.

A further proposal to establish an operational station at Bairnsdale in accordance with the R.A.A.F. Expansion Scheme was approved on October 17, 1941, by the Minister of Air, Arthur Drakeford, subject to scrutiny by the Board of Business Administration at an estimated expenditure of £268,396. This later scheme was for the establishment of a station comprising two General Reconnaissance Squadrons and Station Headquarters.

This would require accommodation for 89 officers, 138 sergeants and above, and 507 other ranks. It was to include all sleeping, messing, ablution, latrine and recreational facilities.

Provision was to be made for splinter-resistant hangarage for 36 aircraft, administrative and workshop facilities, a W/T transmission building, bomb dumps, all engineering services and for aerodrome preparation, drainage and the construction of surfaced runways.

The site then comprised a total area of 690 acres, of which 495 acres were already owned by the Commonwealth of Australia. The additional area of 195 acres required for extension of the site was privately owned land subject to acquisition by the Commonwealth as being essential to the defence of Australia.

BAIRNSDALE BECOMES OPERATIONAL

As early as 1940, the airfield at Bairnsdale was used by the R.A.A.F. although a great deal of timber had still to be removed from the close proximity of the dusty gravelled runways. In January of that year, 'A' Flight of No. 2 (General Reconnaissance) Squadron, then based at R.A.A.F. Laverton was instructed to inspect the aerodromes at Bairnsdale, Pat's River on Flinders Island, Western Junction and Hobart. In addition, the squadron was ordered to carry out parallel track searches from Bairnsdale to Pat's River and also from the northern coast of Tasmania back to Laverton. On January 15, 1940, an Anson of No. 2 Squadron, piloted by P/O P.H.R. Hodge landed at Bairnsdale and the party stayed for two or three days. Hodge was killed almost exactly two years later when his Hudson was shot down by Japanese Zeros while attacking the Japanese invasion fleet near the Celebes, on January 12, 1942.

Anson aircraft of No. 2 Squadron used the Bairnsdale aerodrome for many months as an Advanced Landing Ground. Anti-submarine patrols, diverging searches and clearing searches for the protection of shipping along the coastline were the main functions of this Unit.

In July, 1940, No. 2 Squadron began re-arming with the new Lockheed Hudson aircraft and the same duties were continued right up until the squadron commenced moving from Laverton to Darwin on December 5, 1941.

One of the more notable duties of No. 2 Squadron, connected with the airfield at Bairnsdale, was when it was called upon to safeguard the convoy transporting the second echelon of the Second A.I.F., against enemy vessels or mines, by carrying out security patrols ahead of the ships as they proceeded through Bass Strait. This operation took place on May 1, 1940.

A high-wing Tugan Gannet survey aircraft was based at Bairnsdale during 1940 and for several months this was a familiar sight tracking back and forth over East Gippsland as it conducted a photographic survey of the region.

The first complete unit of the R.A.A.F. to be stationed at Bairnsdale was No. 7 (General Reconnaissance) Squadron equipped with Lockheed Hudson aircraft.

This squadron was formed at Laverton on June 27, 1940, with F/Lt. E.D. Scott as its first C.O. For many months No. 7 Squadron had no aircraft of its own. It was intended to operate Hudsons and both aircraft and personnel from No. 24 Squadron were attached to provide assistance to 7 Squadron in conversion training and instruction on Hudson maintenance. These few aircraft of No. 24 Squadron were returned to Townsville during October, 1940, and by December, as No. 7 was still without aircraft, it was attached to No. 2 Squadron, also based at Laverton, so that the unit's personnel could obtain operational experience on Hudsons. For the ensuing year, 7 Squadron had virtually no separate identity. The officers were posted away and only a small nucleus of ground staff remained.

In January, 1942, 7 Squadron once again became an independent unit. It reformed at Laverton as a Hudson squadron, and during the same month the role of the squadron was temporarily altered to that of an Operational Training Flight of an O.T.U. The first course, consisting of six complete crews and ten extra W.A.G.'s arrived on squadron on February 2, 1942. These courses continued until June, 1942.

As well as performing its operational training duties, No. 7 Squadron was required to carry out seaward searches. Coastal patrols in Australian waters now had taken on a greater significance with the entry of Japan into the war.

The squadron commenced movement to Bairnsdale on April 2, 1942, when an advance party left Laverton. This was followed by a second ground party and on May 1, 1942, by an air party of six Hudsons. The rear party arrived at Bairnsdale on May 5, 1942.

Wing Commander S. de B. Griffith, A.F.C., took over command of the squadron on April 22, 1942. He had been Station Commander at Darwin and had experienced the very heavy raid of February 19, 1942, when losses had been particularly severe.

For most of the month of May, 1942, the squadron was mainly occupied in the move to, and the consequent settling in, of the unit at Bairnsdale. At that time, accommodation on the airfield was meagre with the contractors, A.E. Watts, working hard to provide the necessary hangarage and living quarters for the men. Many local builders were among the force of men engaged in this work and the surviving splinter-resistant hangars which still exist in near-perfect condition after the passage of four decades are testimony to their workmanship.

Because of the severe shortage of accommodation, most of the commercial accommodation in Bairnsdale was taken over by the Air Force. The Victoria Hotel was completely taken over for Officers and the the Main and Terminus Hotels provided billets for Sergeants and Warrant Officers. Transport section was based at the Club Hotel with the transport drivers billeted in the Lounge and their vehicles parked in the street outside the hotel. Other hotels also had their complement of Air Force personnel. One or two of the hotels provided cooking facilities but most of the men had their meals at the aerodrome and came back to the hotels after the evening meal.

The unit also maintained a detached flight at Mallacoota.

By June, 1942, the squadron was once again fully engaged in operational flights off the coast, escorting convoys and mounting anti-submarine patrols.

On June 4 a squadron Hudson was executing a creeping line ahead patrol with the object of locating possible surface vessels and submarines. This aircraft was piloted by F/Lt. C. Williams and navigated by P/O H. Cameron. The crew were well aware of the presence of Japanese submarines off the coast as Sydney Harbour had been attacked only five nights before, and reports of several ships being torpedoed south of Sydney had been received.

They were approaching a British ship, the 'Iron Crown' of 3,353 tons at a position 44 miles S.S.W. of Gabo Island when the vessel blew up in front of their eyes. Soon afterwards a submarine surfaced some distance away and Williams immediately dived on the target, dropping two anti-submarine bombs which were seen to straddle the conning tower. Williams now only had two general purpose bombs left to complete the kill, so he climbed rapidly in order to drop these from a minimum safe height of 500 ft. Turning to run in for the second attack, the crew observed the submarine succeed in submerging and no further contact was made. It was thought that it had been damaged by the initial attack as it had taken some time to crashdive as the Hudson desperately manoeuvred to continue the attack. The crew searched fruitlessly for some time and another Hudson of 7 Squadron was despatched to locate survivors of the stricken ship and drop life jackets. This was recalled as the survivors had been rescued.

With this very vivid evidence that enemy submarines were operating off the south coast of Australia, No. 7 Squadron stepped up its anti-submarine operations. It was during this hectic period that the squadron experienced their first fatal crash from Bairnsdale, when a Hudson captained by S/Ldr. J.W. McGilvray, while returning from a seawards patrol, crashed and burnt at Long Point near Sale, destroying the aircraft and killing the entire crew.

On June 12, Squadron Leader J.A. Hepburn assumed temporary command of the squadron and on June 17, No. 7 Squadron once again lost its identity as it was absorbed into No. 1 Operational Training Unit which had begun its move to Bairnsdale. By June 22 only a few members of the squadron remained under Squadron Leader A.D. Henderson who took over command on that day. Most of the personnel and the aircraft and equipment now formed the Hudson Squadron of No. 1 O.T.U. as 7 Squadron's role as an Operational Training Unit had ceased.

There was no further operational flying during July and August. The squadron, or what was left of it, was once again on the move, this time to Nowra, N.S.W. to convert to Beaufort aircraft. On August 16 the C.O., the Adjutant and the remaining 13 airmen left Bairnsdale for their new home.

Although No. 7 Squadron had served for only a comparatively short time at Bairnsdale, association with that station was to continue over the remaining war years, as many crews posted to No. 7 (Beaufort) Squadron, were trained at Bairnsdale or nearby East Sale.

BAIRNSDALE IN WARTIME:

Bairnsdale had suddenly become an Air Force town. With the hotels and guest houses crowded with airmen while the camp was being completed, the dark blue Air Force uniform became a familiar sight. A number of airmen managed to find accommodation for their wives and families. Others who came initially to Bairnsdale as R.A.A.F. members, married into local families and many of them returned after the war to settle in the Bairnsdale district.

With the town on a war footing, Air Raid Precautions came into force. After the assault on Darwin, volunteers dug hundreds of feet of slit trenches in the grounds of the Primary School and the Hospital.

The Volunteer Defence Corps was formed mainly from the ranks of the "old Diggers" of W.W.1. While their prime task, in case of invasion, was to assist in the defence of the region; in the immediate district they co-operated with the Air Force by sometimes mounting guard at the numerous small R.A.A.F. establishments, radar posts, bomb dumps, etc. in the area.

Early in May, 1942, a two-storied building in Riverine Street, near the old swimming pool, was secured as a club for servicemen and servicewomen during their off duty hours in the town. This was administered by a committee jointly drawn from local women's and Church organisations.

A Services Auxiliary was formed during October, 1942, known as the 'Links of Service', which catered for the welfare of the R.A.A.F. and W.A.A.A.F. members. Part of its function was to assist in running a coffee lounge in the recreation hut at the aerodrome under the supervision of the Red Shield Officer.

Air-minded youth of the town welcomed the formation of a Bairnsdale Flight of the Air Training Corps in September, 1942. This had an initial enrolment of twenty local boys ranging in age from 16 to 18 years. Parades were held each Sunday at the aerodrome when the cadets were introduced to service life and quite often given the opportunity of flying in service aircraft. They were required to attend regular evening lectures, held twice weekly in the High School, where they were instructed in Maths., Science, Morse Code, Aircraft Recognition, Trade Theory, Drill and P.T. This initial training was of considerable benefit to the many A.T.C. Cadets who enlisted in the Air Force on reaching 18 years.

Perhaps the most valuable ancillary service organised among the civilians was the formation of the Volunteer Air Observer Corps. Formed on the same lines as the Royal Observer Corps in Britain, V.A.O.C. members were trained in the speedy recognition of aircraft, and they manned their observation posts twenty-four hours per day to spot and report any planes flying within their vicinity. They worked in close co-operation with the telephone operators who were instructed to give instant priority to 'Air Flash' calls for the transmission of V.A.O.C. reports to the Control Room at Bairnsdale. Speed was vital due to the high speed of aircraft, as a plane's position could change rapidly in a matter of minutes. The V.A.O.C. was to prove of great value in the tracking and search for lost aeroplanes from Bairnsdale and neighbouring airfields; luckily they were never required to give warning of the approach of enemy aircraft.

The V.A.O.C. first came into prominence with a report in the 'Bairnsdale Advertiser' of April 16, 1943, which told the story of how a young R.A.A.F. sergeant-pilot owed his life to the V.A.O.C. after his plane had plunged into a lake. Although the name of the locality was suppressed due to wartime security, the people concerned were readily identified as residents of Paynesville.

Mrs P. James witnessed the accident from her home and after alerting the local V.A.O.C. post, she and her husband rowed out to the wreck where they were joined by Messrs. W. Robinson and L. Greer who had been fishing nearby. The party managed to free the unconscious pilot just as the water was on a level with his mouth. They searched for other occupants but were relieved to find that the pilot had been flying solo. Other members of the V.A.O.C. were on the jetty to administer first-aid while awaiting arrival of the ambulance.

Paynesville had the distinction of being the oldest established V.A.O.C. post in Australia, and the tall wooden lookout tower overlooking the lakes was manned constantly over the war years.

TRAINING PRESSURES INCREASE

With the movement of No. 1 O.T.U. to Bairnsdale, the station commenced its busiest period.

No. 1 Operational Training Unit had formed on December 8, 1941 (the first day of the Pacific War) at R.A.A.F. Nhill on a temporary location basis. Its original equipment consisted of a variety of Fairey Battle, Tiger Moth, Wirraway, Oxford, Beaufort and Lockheed Hudson aircraft and their function was to provide operational flying and instruction to crews preparing to go into combat. It was, in effect, the last step before the aircrews were posted to the fighting squadrons.

On February 14, 1942, all the unit's aircraft together with instructors and pupils departed Nhill aerodrome enroute for West Sale. A special troop train transporting the complete personnel with stores and equipment left Nhill on the 15th and arrived at the new station on the next day, February 16, 1942.

The war situation at that stage was very serious indeed and the unit, in addition to its operational flying training duties, had also to carry out armed seaward reconnaissance in support of the coastal shipping.

From West Sale, an Advance Party of the unit commenced the movement to Bairnsdale on June 14, 1942. The main body of the personnel with stores and equipment completed the journey to their new base on June 17, 1942. They arrived by special train which detrained the personnel at the level crossing known as the B.R.C. crossing about two miles to the west of the town. From there the airmen marched to the airfield.

The new unit was equipped mainly with Bristol Beaufort bombers made in Australia by the Department of Aircraft Production. They also had some Lockheed Hudsons and used Fairey Battles as target-towing aircraft and Airspeed Oxfords as advanced trainers.

Group Captain C.D. Candy, O.B.E., became Commanding Officer of No. 1 O.T.U. He had recently returned from the United Kingdom where he, along with several other Australian Regular Air Force officers, had been attached to the Royal Air Force. He had commanded No. 206 (Hudson) Squadron of Coastal Command, R.A.F. This unit had operated in a convoy protection and anti-submarine role from bases in south-west Britain. After a tour with No. 206 Squadron, Group Captain Candy had formed and been in command of No. 6 (Coastal) O.T.U. of the R.A.F. He was well qualified to command Australia's first O.T.U. specialising in maritime operations.

The pattern of flying begun at West Sale continued at Bairnsdale with operational anti-submarine and convoy protection patrols flown in addition to the more normal operational training exercises.

Initially these armed operational flights were mainly carried out by standby crews drawn from members of the staff of very experienced aircrew employed as instructors. These comprised not only permanent Air Force personnel, but included many of Australia's most experienced commercial airmen who had accumulated thousands of hours of civil flying in the hazardous pre-war days.

Prior to arriving at an O.T.U., the Pilots, Navigators (then called Observers), Wireless Operator/Air Gunners, had confined their training to schools established for their individual musterings. Here at 1 O.T.U. they 'crewed up', the crew consisting of one pilot, an observer and two W.A.G.'s. The pilot was usually the captain of the aeroplane, irrespective of his actual rank while on the ground. He was responsible not only for the safe operation of the aircraft, but also he had to ensure that the plane was correctly loaded with fuel and bombs for the designated mission. He also had to work out suitable tactics to be undertaken in the event of a 'strike' or if evasive action was required. The observer had the job of not only navigating the aircraft to the target and return, but was also the bomb aimer, coding expert, weather recorder, shipping identification expert and was required to have a working knowledge of flag and Aldis signals as practised by shipping. The two W.A.G.'s were responsible for the guns, turrets, supplies of ammunition and the operation of the radio using the specified frequencies of the day and later the operation of radar and other specialised equipment.

After the 'crewing up' process, the pilots spent some time in converting to the newer aircraft before the complete crews began intensive training by instructors, most of whom had experienced considerable operational flying against the enemy. The type of aircraft employed by the pupils at O.T.U. would be of the same type on which they would fly on operations in the very near future. The crews themselves would most likely stay together from O.T.U. right through their tour of active service.

Student crews were expected to master a complex series of air exercises involving navigation, photography, bombing and gunnery tactics, shipping recognition, reconnaissance, signalling and searchlight evasion, etc. In fact, they were trained to meet and cope with any problems which they could meet in actual operational flying.

Crews had to operate their aircraft with full military load of fuel, bombs and ammunition and fly in all sorts of weather. The transition from the gentle Avro Ansons or Oxfords to the much more spirited Hudsons, and the Beauforts in particular, was a very testing time for all the crew, but especially for the pilots.

The exercises involved long flights out over the sea of sometimes up to 7 or 8 hours duration. One exercise flown was down past Eddystone Light on the east coast of Tasmania and then on to a south-easterly course completing various navigation exercises occupying about three flying hours before returning to Tasmania for the navigator to get a land fall to calculate a course back to base. It was always hazardous work flying from Bairnsdale out over Bass Strait at very low altitudes to simulate radar evasion. The weather conditions were usually adverse and were said to be similar to those experienced in the Bay of Biscay. This was particularly dangerous for the inexperienced crews as they were sometimes required to fly at night faced with the added hazards of rain, icing and notoriously poor radio reception due to the phenomenon known as 'skip distance' where there is poor reflection of wireless signals from the earth or the ionosphere. This required good technique on the part of the Wireless Operator to select the correct frequency, length of aerial, etc. to keep in contact with base.

Fog was also very prevalent at Bairnsdale during the winter months of 1942. When fog rolled down from the foothills and covered the coastal plain in those days, there was no equipment available to guide the Beauforts and Hudsons down to the flare path, so the only choice for the tired crews was to divert to either Mallacoota or Sale. At that time neither were ideal dromes to attempt a landing on a dark, wintry night.

Training was greatly accelerated during the dark days of 1942 when the war news was at its most serious. Crews were urgently needed 'up North' to stem the tide of Japanese advances.

All these factors led to a very high accident rate. At one stage aircrew losses at Bairnsdale were said to be higher in comparison than losses suffered in action against the enemy.

HIGH ACCIDENT RATE

The first crash occurred on July 5, 1942, when a Beaufort dived into the ground while on final approach to the Bairnsdale aerodrome. On August 9 two airmen were killed when a Hudson crashed and burnt after diving into the ground from about 2,000 ft. August 27 was a black day. Two Beauforts flying in formation about five miles from the airfield, in the vicinity of Hillside, collided in mid-air. One Beaufort crashed, exploding on impact, killing its crew of four, the other barely made it back to the aerodrome. That same day another Beaufort swung violently on landing and ground looped with severe damage.

Accidents continued in September with a Beaufort damaged in a landing accident while on a flight to Essendon on the 9th. Another Beaufort (A9-28) failed to return from a seawards reconnaissance patrol on September 12 with the loss of its crew of four. A Hudson taking off on September 19 ground looped, collapsing the undercarriage, when the starboard engine failed, and on the night of September 29, a Beaufort, attempting to land at Bairnsdale, overshot the flare path, and crashed through the boundary fence.

October 1942 was the worst month for accidents while 1 O.T.U. was stationed at Bairnsdale. On the night of the 2nd a Beaufort (A9-73) was destroyed when it crashed and burnt on takeoff, killing its pilot, the sole occupant. That same night, Beaufort A9-49 flew out to sea and was not seen again, losing another crew of four. A little over a week later, on the 11th, Hudson A16-18 also disappeared over Bass Strait with its crew while enroute from Bairnsdale to Flinders Island, and on October 21 the crew of Beaufort A9-86 were fortunate to survive a forced landing of their aeroplane when both engines failed. Beaufort A9-57 was damaged on October 26 when its undercarriage was retracted while it was still on the ground, but the following day saw the most tragic, if spectacular, crash of all.

A public relations visit had been arranged for Wednesday, October 27, and representatives of the press and the newsreels were on hand to record some of the activities of this 'training unit somewhere in Australia'. The climax of the day was to be a simulated diving attack on the airfield, demonstrated by A16-38, a Hudson captained by F/Lt. Frank Tampion. On pullout at a very low altitude, the aircraft lost the complete starboard wing and performed 1¾ rolls before plunging into the centre of the airfield and exploding. As Tampion and his crew of three had no chance of bailing out, they were killed instantly. There a legendary story of how one wheel of the plane travelled for many hundreds of yards from the crash site.

The student crews at this time were being affected by the high accident rate, and the C.O. had to take strident steps to maintain morale, reduce accidents and continue to meet the intense training programme. In an effort to increase safety, any breaches of air discipline were dealt with severely. Social events were organised in an attempt to relieve the tension. These included an Air Force Athletic Carnival held at the Bairnsdale Showgrounds on October 3 which attracted a large crowd and the first R.A.A.F. Ball was held in the Prince Regent Theatre (now Cremin Hall) during November.

Another, perhaps more practical move, took place when the C.O. gave permission for a Beaufort to be stripped of any movable equipment. The aircraft's flying ability was then amply demonstrated by F/Lt. C. Williams. This is remembered as an amazing display of aerobatics in an aircraft that many had begun to look upon as a 'hoodoo' machine. Eye-witnesses recall that he looped and rolled the Beaufort and at times turned so fast that vortices began to form at the wingtips, a phenomenon rarely seen in those days. It was a great morale booster for the student crews, as it pointed out that a Beaufort in the hands of a capable pilot possessed a sparkling performance for an aircraft of its size and type.

November saw a marked reduction in the casualty rate. Three crashes, all involving Beauforts occurred, the first on November 14 when A9-72 ground looped after landing. The next day, Beaufort A9-100 was scrapped after the starboard engine had failed at 500 ft. and on landing on one motor at Pat's River, Flinders Island, it crashed through a fence. An officer and three airmen were killed on November 17 when Beaufort A9-45 crashed on a non-operational navigational exercise about five miles from the Bairnsdale aerodrome.

WOMEN IN UNIFORM

The first members of the Women's Auxiliary Australian Air Force to arrive at R.A.A.F. Bairnsdale were probably two W.A.A.A.F. medical orderlies who came to the station on October 8, 1942. As more accommodation became available, scores more W.A.A.A.F. personnel arrived to take their places working beside the airmen in almost every trade. The W.A.A.A.F's served as telephonists, telegraphists, transport drivers, parachute packers, fabric workers, general clerks, cooks, stewards, aircraft riggers and cleaners, cypher clerks, dental and postal orderlies, instrument repairers and assistants in the meteorological section. They worked the same hours, wore the same shapeless blue overalls and shared the same conditions as the men. Soon airmen in every section became accustomed to the presence of airwomen working alongside them and they came to be accepted as equal members of the Air Force. Their contribution to the war effort was enormous.

Their initial integration into the running of the Station, however, was not completely without problems. In one respect the R.A.A.F. was not ready for the influx of servicewomen and that was in the provision of latrines. Service discipline dictated that the system of rank extended even to the more basic functions, and all latrines were clearly marked for the use only of 'Officers', 'Sergeants and W.O.'s', and of course, 'Airmen'. No one had thought of 'W.A.A.A.F's'.

The first batch of airwomen to work on tarmac duties found this out to their great discomfort. By mid-morning on that first day, the party was becoming quite ill at ease. Their immediate superior, a male sergeant, although in sympathy with their plight dare not offend convention, but he, in an inspired moment, offered the loan of a small tractor pulling a bomb trolley to transport the now distressed females back to the Waafery. Selecting one as a driver, he started up the tractor and stood back as they proceeded at a fast rate of knots in the direction of the camp.

Unfortunately, a group of young officers on course, chose that moment to march down the same road towards the tarmac. While the girl driving knew how to steer the machine, she had not taken the time to find out how to stop it. The officers scattered for their lives as W.A.A.A.F.'s, tractor and trolley tore through their ranks. By then, of course, it was too late. The shrieks of laughter at the sight of those officers taking violent evasive action had proved far too much for them.

TO THE BATTLE FOR BUNA

An operation unique in the annals of the R.A.A.F. took place in the months of December, 1942, and January, 1943, when a flight of 15 Hudson aircraft left Bairnsdale and flew over two thousand miles to assist the front-line troops then engaged in the Battle for Buna in Papua.

Word was received at Bairnsdale on the night of December 10 that the situation was desperate due to a shortage of supply aircraft. The Japanese had been pushed back to the coast and vicious fighting had been going on since mid-November to dislodge them from their positions around Buna, Gona and Sananandа. American and Australian engineers had succeeded in preparing advanced bases at Popondetta and Dobudura to enable much needed supplies to be flown in across the Owen Stanleys from Port Moresby. However an acute shortage of suitable transport aircraft had placed the whole operation in jeopardy. A special R.A.A.F. Transport Flight was hurriedly organised using a motley collection of planes from both service and civil sources in Australia. These included the Hudson Squadron of 15 aircraft from 1 O.T.U., Bairnsdale, and they were crewed by the experienced instructional staff, all of whom had previous combat experience in the opening days of the Pacific war. Crews training on Hudsons at Bairnsdale were sent on leave subject to recall on short notice and instructors not included in this special unit were temporarily transferred to help train Beaufort crews. The entire unit was under the command of S/Ldr. O.B. Hall, D.F.C., A.F.C., a very experienced pre-war commercial pilot. The Bairnsdale contingent was known as No. 1 O.T.U. Detached Flight.

Mr Bert Rodd, then a Pilot Officer observer instructing at 1 O.T.U., remembers being woken up at night and being briefed on the forthcoming operation. The next morning, December 11, was taken up with last minute preparations and the flight of 15 Hudsons took off at 3 p.m. on their long, gruelling flight to Papua. They flew far into the night, refuelling at Parkes and Archerfield. They continued flying all the next day in an effort to get within easy range of Port Moresby to enable the formation to slip in to Ward's Strip early on the morning of December 13 before the first raid of the day, for at that time Port Moresby and airfields were under constant seige by the Japanese Air Force.

The flight up north was not without incident. Three Hudsons were involved in minor accidents — A16-110 in a taxiing mishap with A16-45 at Parkes, and A16-11 was damaged at Archerfield when a Lodestar of the U.S. Army collided with it. The most hair-raising incident befell the crew of A16-26 which landed at the civil aerodrome Cooktown, damaged, after spinning 1,000 ft. during a severe tropical storm.

The R.A.A.F. Special Transport Flight began operations from Ward's

Strip on December 14 with a supply drop at Soputa. As the distance between Port Moresby and the Buna Front was comparatively short, about 100 miles, average flying time was approximately 35 minutes. This enabled an aircraft, with improved ground handling, to put up as many as four or even five sorties each day, depending on the weather. The terrain, like much of Papua/New Guinea was extremely mountainous and the 'Gap' in the Owen Stanleys through which the planes passed would close down each day about 3 p.m. Top cover was provided by fighters to give the transports protection from enemy attack and the transports themselves, once they penetrated the 'Gap', flew as low as possible just skimming the tops of the jungle. If any aircraft failed to make their way back to Port Moresby before cloud closed the 'Gap', it was necessary to fly right down to Milne Bay, then across to Kila Kila in order to return to Ward's Strip.

It was on one of these supply drops over Soputa that the unit lost their first crew on only the second day of operations, December 15.

In answer to an urgent call for an ammunition drop by an A.I.F. unit fighting in the jungle near Soputa, three Hudsons, captained by S/Ldr. Gibbes, F/Lt. Landrey R.A.F. and S/Ldr. W.A. Pedrina, took off from Ward's Strip early in the afternoon. Pedrina's crew had already made three supply drops during the morning and although the 'Gap' by this time was closed by cloud, they managed to pick their way through at 6,500 ft. before dropping down to tree-top level.

The three planes were orbiting after making two circuits dropping supplies from about 200 ft., when, on its final run, Pedrina's Hudson A16-36 was seen to go into a steep climb and then explode in mid-air. The pilot and two of the crew were killed, but the turret gunner, F/Sgt. L. 'Wonga' Callagahan fell into a palm tree and survived the crash although seriously injured, sustaining a broken pelvis and forearm. He was rescued by ground troops, but couldn't recall much except a tremendous explosion, then found himself hanging in the trees. He had not parachuted but had been blown clear of the aircraft at practically tree-top level.

It was never established for certain how this aircraft was shot down. It was held by some that a sniper in the tree-tops had shot the pilot, while many suggest that it was a direct hit from an anti-aircraft gun operated by the Japanese from the infamous 'Gona wreck' situated nearly 7 miles away. This was a wrecked ship lying off Gona on which the Japanese had established a crack anti-aircraft gun crew who had a reputation for extreme accuracy.

The loss of Pedrina and most of his crew was felt keenly by the other members of the Flight. He had been a very popular and well respected pilot, in a unit which contained many distinguished pilots, where he had been known as"Pedro, the Perfect Pilot", not without justification.

This was a very hectic period for the crews and the group of key maintenance men who had accompanied the planes from Bairnsdale. The latter were kept working around the clock back at Ward's Strip, keeping the aircraft flying in spite of a succession of enemy bombing and strafing attacks. The aircrews, after those long hours of hazardous flying each day, had little rest at night at the camp as the crash of bombs and the roar of the 'friendly' A/A barrage each night made sleep almost impossible. To add to their discomfort, the operation took place in the middle of the 'wet' season with its oppressive heat and humidity.

On December 26 the Japanese fighters succeeded in penetrating the fighter cover and caught several Hudsons on Dobodura strip. The Hudsons were always left with the motors running, ready for immediate take-off, and the gunners always manned the turrets while on the ground. In this, their first raid on Dobodura, the enemy 'Haps' (a version of the Zero), bombed and strafed along the new strip. F/Lt. Landrey hurriedly took off with five fighters attacking him. He managed to evade their attentions by hugging the terrain and succeeded in getting back to Ward's Strip. His gunner, Sgt. Ross, was struck by a bullet which creased his scalp. Landrey's Hudson made a safe landing although its tyres were punctured by bullets and minor damage done to the airframe. Ross had scored several hits on one fighter but the results were unobserved.

The C.O., S/Ldr. Hall, was also attacked on take-off by about seven fighters. He also managed to get his plane back to Ward's Strip although in a badly damaged condition. The crew were unhurt and his gunner had shot down one fighter, probably destroyed another and damaged a third. An incendiary bullet caused a fire in the Hudson, which the wireless operator extinguished.

F/Lt. Marshall managed to take off under attack, evading the enemy by flying at low level. He returned to the strip and landed again with his gunner still firing at the fighters after landing.

The crew of the fourth aircraft were not so lucky. S/Ldr. Hemsworth and his crew were engaged in loading A.I.F. wounded into their Hudson when the fighters struck. They had two wounded men in a crate-like stretcher and two walking wounded aboard and were about to take on others when the raid began. Hemsworth weighed up his chances and decided to take off rather than risk being shot up on the ground, particularly as he was carrying Australian wounded. He took off with three Japanese chasing him and proceeded to hug the trees and by flying up and down the gullies hoped to evade the fighters.

This action was to last for the next 27 minutes, during which the Hudson was under constant attack by at least two fighters. The turret gunner, F/Sgt. Stephens, registered hits on one fighter but soon after this one of the Japanese closed in and put the turret out of action, wounding Stephens very severely. The observer, who all this time had been blazing away with the two waist guns and trying to assist the wounded Diggers, managed to get the gunner out of the turret, get it operating manually and render first-aid to the badly injured Stephens. The wireless operator, F/Sgt. Bamber, who had been vainly calling for fighter protection, took the place of Stephens in the turret and got off a good burst hitting another of the fighters. From then on they received no further attacks.

Noticing a group of Corvettes and landing ships standing off shore near Oro Bay, Hemsworth flew out towards them hoping to put his aircraft under the protection of their guns in case of further attack. On crossing the beach, an American anti-aircraft unit that had just landed, mistook the Hudson for an enemy aircraft and promptly opened fire on it. The oil reservoir operating the automatic pilot directly in front of the pilot was hit by an incendiary bullet and set on fire. Hemsworth was severely burned about the face, chest and arms, and was forced to ditch the Hudson on the sea immediately.

By this time, two of the wounded men had been killed and all of the crew were wounded, Stephens later dying of wounds. The survivors who included two A.I.F. wounded and the crew, managed to climb aboard the dinghy and get clear of the aircraft. Even then their troubles were not over for they found that their dinghy was drifting into shore and soon they found themselves under fire from a Japanese position on Sanananda Point. A crew member slipped over the side, trying to tow the dinghy away by swimming, but found it impossible due to his injuries. Then an American P.T. boat came roaring up, dropped a line and towed the dinghy to safety as machine-gun bullets from the shore were raising spouts in the water.

The wounded were put into hospital in Oro Bay were Stephens died. Here, the rest of the crew were happy to be reunited with 'Wonga' Callagahan whom they had believed to have died in the crash of Pedrina's Hudson. Later they were moved to another hospital in Milne Bay by R.A.N. Corvette and finally were flown down to Townsville.

Bert Rodd was the only member of that crew to return to 1 O.T.U. at Bairnsdale many weeks later in March, 1943. The other two crew members survived, but spent further time in hospital before posting to other units. All three survived the war.

The Special Transport Flight continued operating until January 11, 1943. On New Year's Day, 1943, bad weather forced four of the Hudsons to stay at Dobodura overnight and on that day, A16-105 was flown back to Townsville from Ward's Strip for repairs. On January 5, General Blamey was flown in Hudson A16-34 captained by the C.O., S/Ldr. Hall, from Ward's Strip to the Buna front. After a stay of two days, Blamey was returned in the same aircraft to Ward's Strip. There were a few anxious moments on the return flight as word was received just prior to take-off that six enemy fighters were approaching the strip. Fortunately the Hudson was well clear of Dobodura before the enemy formation arrived. By this time, the unit was short of pilots, so on January 8 the C.O. of No. 36 (Transport) Squadron, S/Ldr. Harry Purvis, an old O.T.U. hand, helped out by flying one of the Hudsons in a delivery flight of food and troops to Bena Bena and also to Wau under a strong fighter cover. On January 9 F/Lt. Dunne, D.F.C., left Ward's Strip for Wau to bring out urgent medical cases but bad weather prevented his landing at Wau. However, the next day, January 10, six aircraft ferried troops and food to that airstrip.

The Flight received orders to return to the mainland on January 11, 1943. From arrival on December 14 until operations ceased, the unit had made 645 flights over the Owen Stanleys, carried in 1,557,318 lbs. of much needed supplies and ferried 1,107 troops during a total of 1,020 operational hours. From Dobodura alone the transport aircraft had evacuated an average of more than 100 casualities, sick and wounded, per day during December and early January.

The remaining few hours at Ward's Strip found the ground staff hard at work, replacing bomb racks and other gear that had been removed, to prepare the aircraft for the return flight to Australia on January 12.

Leaving Ward's Strip at 0725 hours on January 12, 1943, six Hudsons flew to Townsville via Iron Range and the other five flew direct to Townsville. All

Lockheed Hudson A16-89 of No. 7 Squadron after a forced landing early in 1942. On August 9, 1942, while on the strength of No. 1 O.T.U., A16-89 crashed and burned at Bairnsdale.

Proving that it could be done, Beaufort T9540 flies on one engine. First of a batch built in Australia for the R.A.F., it was taken over by the R.A.A.F. and re-numbered as A9-1. It served with No. 1 O.T.U. until it overshot the flare path and crashed through the boundary fence at Bairnsdale on the night of September 29, 1942.

ABOVE: Lockheed Hudson, A-16-34, piloted by S/Ldr. Hall, C.O. of the 1 O.T.U. Special Transport Flight, flew General Blamey from Port Moresby to Buna on January 5, 1943. BELOW: A16-3, another Hudson of the Special Transport Flight, piloted by F/Lt. Hemsworth, was shot down in error by an American Anti-Aircraft Unit into Horiko Bay on December 26, 1942.

Bairnsdale Airfield.
Taken in 1980, this aerial view shows the system of runways and taxiways which date from the war. Also visible on the original photograph are the foundations of many of the war-time buildings.

LEAVING FOR WORK: A typical scene early each morning in Bairnsdale as R.A.A.F. tenders collected parties of airmen from their billets in local hotels while accommodation at the airfield was still under construction.

Essential to the survival of aircrew forced down in the sea was the correct operation of rubber dinghies carried in the aircraft. An aircrew practices dinghy drill on the peaceful Mitchell River, Bairnsdale, in 1942.

Remains of one of the many Beauforts which crashed in the Bairnsdale area during wartime. A9-143 of No. 1 O.T.U. crashed and exploded about four miles from Paynesville on July 30, 1945, killing the crew of four.

eleven planes reported safely at Amberley at 0915. F/Lt. Gibbes and S/Ldr Hall flew on to Richmond arriving at 2305. The next day, January 13, the remainder of the flight flew into Richmond arriving at 1100 hours, and after 24 hours rest, the entire formation left Richmond at 1100 hours and arrived back at Bairnsdale at 1345 hours on January 14, 1943.

Their superb effort at a critical time in the Battle for Buna and Gona contributed a great deal towards the success of this, the first major defeat of the Japanese land forces in the South Pacific. The returning instructional staff brought back up-to-the-minute experience in tactics and procedures to pass on to the courses. In their absence, training during December had been kept to the minimum and no casualties had occurred. However, with the New Year, training was once again in full swing and casualties rose proportionately.

TRAINING LOSSES CONTINUE

Beaufort A9-147 was landing after an operational exercise on January 7, 1943, when the wind changed rapidly through 180 degrees and continued blowing at 15 m.p.h. The aircraft at that time was on finals at about 200 ft. and the pilot made a very fast landing, running through the end of the runway, and ground looped. The crew were fortunate to escape injury as the aircraft was extensively damaged.

Another Beaufort, A9-160, lost brake pressure on landing on January 11, and ground looped at high speed to the left, wiping off the undercarriage and tail wheel. Once again the aircraft was badly damaged but the crew members suffered only minor abrasions.

That same day, however, the first fatalities for the New Year occurred when Beaufort A9-67 failed to return from a seaward navigation reconnaissance exercise. The crew, captained by Sgt. W.J. Long, were presumed lost.

On the night of January 21, the pilot of A9-121 returning from a night exercise, stalled his aircraft on landing at approx. 30 ft. The port wing dropped and the aircraft landed on that wing and the port undercart. The resultant damage was very severe, both engines being torn out, with the crew extremely lucky to escape before the wreckage caught fire and was totally destroyed.

A veteran Fairey Battle, on drogue towing duties for air to air gunnery practice, was the next to be involved in an accident, on January 28. It was flying at 800 feet about 2½ miles N.W. of the drome when the engine blew smoke and cut out. The pilot selected a good field for landing but found H.T. wires to the leeward. He was forced to make a short landing, overshot and crashed through the fence. The pilot was unhurt and the drogue operator sustained minor abrasions to the scalp and knee but the Battle was very badly damaged.

That same day, the crash of an Oxford BG473 on airmanship exercises when flying about four or five miles from the drome took the lives of two young pilots and a fitter who was a passenger. It was seen to commence a spiral turn then slowly spin into the ground where it was totally destroyed.

At the end of January, 1943, No. 1 O.T.U. at Bairnsdale had on strength 235 officers, 420 sergeants and above, and 1,312 other ranks. There were also two W.A.A.A.F. officers, one W.A.A.A.F. sergeant and 114 airwomen, the total personnel being 2,114. In addition, the number of pupils in training numbered 285, and the progressive number of pupils who had graduated since the commencement of training at Bairnsdale numbered 333.

There were exactly 100 planes on strength comprising 50 Beauforts, 12 Hudsons, 25 Oxfords, 12 Battles and 1 Tiger Moth. These aircraft had notched up a total of 3,321 flying hours during the month of January of which 2,121 hours were spent on Beauforts.

January and February, 1943, saw a resumption of Japanese submarine attacks along the south coast of N.S.W. and anti-submarine and convoy escort duties were included in the operational exercises for the trainees to gain experience in Naval co-operation and ship recognition. There were reports of Japanese submarines in Bass Strait and several attacks were made by aircraft from Bairnsdale. Apart from decimating the whale population in those waters, no positive results were observed. Perhaps the continued presence of Beauforts and Hudsons made life a little more difficult for the Japanese submariners.

Some very interesting ships were spotted during these flights including the 'Mauretania', the 'Queen Elizabeth' and the 'Ile de France', a speedy, four-funnelled, luxury liner. All of these were acting as troopships. One of the most picturesque ships seen was the 'Lawhill', a barquentine, magnificent under full sail. This was one of the last of the legendary grain ships.

It was an impressive sight for the crews to witness a large convoy passing through Bass Strait, continually ziz-zagging with their flanks guarded by Naval escort vessels. The threat of a German raider raised an alarm when it was reported to have been spotted by an American vessel well to the east of Gabo Island. This sent aircraft hurrying out to sea for hundreds of miles on an easterly course with no results. Operations such as this were usually undertaken by crews taken from the more experienced instructional staff.

Early in February, 1943, the local press announced that rehearsals had commenced at the Air Force Station for a R.A.A.F. Variety Show called 'Tiggerty Boo'. This stage show, produced by the Adjutant of No. 1 O.T.U., F/Lt. David Munro, was one of the best morale raisers in the unit's history. It originated from a suggestion by the Chaplain, F/Lt. Selwyn Harris. Among the artists participating were many inter-State performers of radio and stage and the sets and lighting were the equal of many a professional city performance. It was originally intended to put it on in the Prince Regent Theatre for a one-night stand only, but the Theatre was completely booked out days prior to the first night and 'Tiggerty Boo' had a memorable season of three nights in Bairnsdale, plus one night in Orbost.

At the station, training was still running at peak with long flying hours to take advantage of the better summer conditions. Unfortunately, aircraft losses still continued. February started badly when a Beaufort, A9-155, crashed shortly after take-off about 1 mile S.W. of the 'drome with the loss of its crew of four. Only one other incident took place in February when a Hudson, A16-102, ground looped on landing.

March saw an increase in aircraft accidents. No less than three separate accidents took place on the 9th. Beaufort A9-6 force-landed about 8 miles north of the airfield; Beaufort A9-41 was written off after crashing in shallow water on Lake Victoria; but the most serious crash occurred that night about 10 miles S.E. of Woodside when Beaufort A9-16 crashed in flames during a night exercise to seawards. Its crew of four died. On March 12 another Beaufort, A9-104, crashed on take-off, and three nights later, A9-81, while practising searchlight evasion with an Army searchlight battery, crashed and exploded in heavy timber about 7 miles from Swan Reach. The pilot was thought to have been dazzled and disorientated by the searchlight beams.

For some time, the majority of crashes had involved mainly Beaufort aircraft. Speculation was rife as to the cause of the accidents. Many theories were advanced — among them pilot error, the use of inferior materials during manufacture, rushing trainees through far too quickly on an aircraft that was difficult to control, poor maintenance and even that of carbon dioxide entering the cockpit, thereby immobilising the crew and allowing the aircraft to crash. The controversy still raged through the press some twenty years after the events, but perhaps the Chief of Air Staff, Air Vice-Marshal Jones summed it up correctly when at a meeting of the Advisory War Council early in May, 1943, he stated that a Court of Enquiry, held into the series of Beaufort crashes at 1 O.T.U., found that "they could not be attributed to any one specific cause".

Many of the crashes were quite inexplicable at the time, with Beauforts suddenly entering a vertical dive from which they did not recover. This situation continued well after the unit left Bairnsdale and operated from East Sale. It was not until early in 1944 that an elevator trim tab malfunction was diagnosed and modified. While Beauforts continued to crash for a number of reasons, the extremely high accident rate was considerably reduced after this modification. Records, however, show that over 90 Beauforts crashed between 1942 and 1945 in East Gippsland and the resultant loss of life was high indeed.

No. 1 Operational Training Unit moved to East Sale in April, 1943. This was to be their final wartime movement which took place officially on April 22, 1943, but not before losing another three Beauforts at Bairnsdale. The bombing range at Bengworden claimed one crew when Beaufort A9-51 crashed and was destroyed by fire during masthead bombing exercises. Five days later A9-153 crashed on take-off and was 'converted to components' or otherwise scrapped. The last crash by an aircraft of No. 1 O.T.U. based at Bairnsdale occurred on April 16 when Beaufort A9-259 also crashed on take-off. This was a brand new aircraft, having been delivered just two days prior to the accident.

Although losses both in men and aircraft had been severe during their stay at Bairnsdale, No. 1 Operational Training Unit had conducted a programme of very intensive training under particularly difficult conditions, coming as they did to a station still under construction in the middle of a severe winter. They had also experienced the teething problems of an aircraft, the Beaufort, quite new to the R.A.A.F.

Under the circumstances existing in those hard 10 months, the unit had passed out 631 pupils and had flown a total of 38,875 flying hours. At one stage it was the proud boast that 1 O.T.U. had an aircraft in the air every minute around the 24 hours.

G.R.S. BRINGS CHANGE OF PACE

Two days after No. 1 Operational Training Unit had officially taken up station at East Sale, the General Reconnaissance School arrived at Bairnsdale from Cressy.

It had originally been planned that this school would be the first unit to occupy the aerodrome at Bairnsdale back in 1940. Due to the tremendous demand for available aircraft to initiate the Empire Air Training Scheme, the establishment of such a school at Bairnsdale had been deferred.

G.R.S. as it was commonly known, had been formed as a separate unit of the R.A.A.F. from the Advanced Training Squadron of No. 1 Flying Training School at Point Cook on April 29, 1940. The unit's strength then consisted of eighteen Avro Ansons. Laverton was the home of G.R.S. until April 20, 1942 when it commenced the movement to Cressy. Officially G.R.S. became established at Cressy on April 27, 1942 although elements of the unit continued to arrive until May 21. Almost one year later the G.R.S. made a final move to Bairnsdale, arriving on April 24, 1943.

G.R.S. was to remain longer at Bairnsdale than any other unit, finally being officially disbanded from that station on March 13, 1946.

The Commanding Officer during 1943 and most of 1944 was Wing Commander D.W. Kingwell, a Regular Air Force officer who had been C.O. of No. 32 (General Reconnaissance-Bomber) Squadron during the first Japanese attacks on Port Moresby in February and March, 1942. He had been injured in action while piloting a Hudson on a photographic sortie over Salamaua in March, 1942.

The main function of the General Reconnaissance School was the training of pilots in general reconnaissance and photography missions and also the training of officers for specialist navigation duties. With the Allied advances through the Pacific Island chain, shipping recognition and naval co-operation were high on the list of requirements for operational crews. The close proximity of Bairnsdale to Bass Strait afforded splendid opportunities to practise reconnaissance and navigational exercises over convoys passing through that stretch of water. Combined with these school duties, the students, like their predecessors of No. 1 O.T.U., kept a lookout for enemy submarines lurking along the coastal shipping routes.

Pilots arriving as trainees to G.R.S. were required to learn much of the art of navigation as part of their studies. That, and the many "pure" navigators who did the G.R. course, necessitated the employment of a large number of staff-pilots who chauffeured the trainees around on their various exercises. On daytime exercises, pilots 'on course' did their own flying, but staff-pilots were always required to accompany them on night flights. On all exercises when the student was a navigator, staff-pilots flew the aircraft. Other duties of the staff-pilots comprised check flights of all incoming pilots, many who may not have flown an Anson for some time, if ever, and testing aircraft after periodic maintenance. There were an equally large number of Wireless Operator/Air Gunners also on staff to provide communications between the aircraft and base, vital in obtaining radio fixes over the trackless ocean on navigational exercises.

Students coming to Bairnsdale during this new phase were not all fresh from training schools. Many were very experienced operational aircrew coming down from front-line squadrons and destined to become specialists in shipping reconnaissance and advanced navigation among other duties within the framework of their particular units.

The instructors were all highly qualified personnel, all with operational experience either in the Pacific, or later in the war, from the overseas theatres.

With the shift from the defensive to the offensive, as the Allied forces gradually wrenched the initiative from the Japanese and began their 'island-hopping' strategy, the need for many of the fighter squadrons had considerably diminished and the emphasis now centred on strike aircraft. These included the low-level types such as the twin-engined Beaufighter and Mosquito. In mid-1944, one course at G.R.S. comprised most of the pilots of No. 86 (Fighter) Squadron, who after operating Kittyhawks from Merauke in Dutch New Guinea, had their unit disbanded and founded themselves, to their chagrin, flying Ansons around Bass Strait. Some of them returned to operations on the heavier ground attack types, but several finished the war still shepherding convoys as pilots of No. 67 Squadron Ansons, based at Laverton.

Ex-students of G.R.S. went to a very wide range of multi-engined squadrons of the R.A.A.F. in the South Pacific. Many aircrew went to nearby East Sale to train on Beauforts; some, on passing out from Bairnsdale, were posted to fly Catalinas, Liberators, Mitchells and other operational aircraft. One batch of navigators were posted on attachment to American bomber units when a shortage of their own trained navigators became evident.

After the departure of No. 1 O.T.U., serious aircraft accidents at Bairnsdale became almost a rarity. This was because the aircraft used by G.R.S. was mainly the Anson, affectionately known as the 'Aggie'. This was regarded by the aircrews as being practically viceless. Accidents did however happen occasionally due to pilot error or the vagaries of the weather. Rough landings accounted for some undercarriages collapsing, but as the Anson landed at a much lower speed than either the Beaufort or Hudson, damage usually was slight in comparison. Some aircraft, due to bad weather or navigational error, crash-landed on the sea, lakes or surrounding countryside, but once again, rarely with serious injuries to the crews.

One such incident took place shortly after the new unit was installed at Bairnsdale. An alert was posted to all district V.A.O.C. posts on the night of May 12, 1943, to keep a lookout for an Anson which was overdue from a flight out to sea. Lights were turned on at several coastal ports to guide the pilot in. Eventually the Anson was sighted by Miss M. Beveridge, a volunteer air observer on duty at Lakes Entrance. It was obvious from the sound of the aircraft that motor trouble had developed. She reported that the plane had disappeared behind the sandhills on the shoreline east of Lakes Entrance. A search was organised by car and boat; shortly afterwards the plane was found to have landed safely on the beach with the crew uninjured.

A little over a week later, Anson W2243 was scrapped after it crashed at Won Wron on May 21, 1943.

Air Force activity of quite another kind was taking place on the Mitchell River over the winter months as crews practised for a R.A.A.F. Regatta rowed on the Mitchell on Saturday, August 14. The Commanding Officer, W/C Kingwell, an enthusiastic oarsman, rowed as No. 2 in the 'Eight'. Crews ranged from novice to the more experienced, and a crew of four W.A.A.A.F's participated. All the racing craft were loaned for the occasion by the Bairnsdale Rowing Club. The Regatta drew more than 1,500 spectators. To keep the event in perspective, an aerobatic display took place over the Regatta by a Ryan Trainer which caused some concern among the crowd as it simulated a forced landing on the river flats opposite the Rowing Club.

Bairnsdale was still being used as a satellite 'drome by aircraft from East Sale. Hudson A16-113 of 1 O.T.U. crashed on take off from Bairnsdale on June 11, 1943 and was extensively damaged. A more serious crash in the district occurred on October 9, 1943, when Beaufort A9-301, while low flying in the vicinity of Lakes Entrance, crashed in the channel near the Kalimna Hotel, killing the crew.

On Saturday, October 16, 1943, hundreds of members of the public were attracted to a rare 'Open Day' staged at the Bairnsdale airfield to view the visiting Avro Lancaster bomber 'G for George'. This was flown into the station by an ace Australian bomber crew captained by F/Lt. Peter Isaacson, D.F.C., A.F.C., D.F.M. It was an impressive sight in those days, as it dwarfed the much smaller aircraft on the tarmac. 'G for George', a veteran of No. 460 (Australian) Squadron, toured Australia to publicise the 4th Liberty Loan. That Lancaster now is a featured exhibit in the Australian War Memorial in Canberra. A female pilot of the British Air Transport Auxiliary, Flight Officer M. Gething, accompanied the Lancaster crew. She gave an exhibition of aerobatics in one of the station's Tiger Moths.

As Japanese submarine activity had continued along the east coast during the period up to July, 1943, aircraft of No. 67 Squadron flying Ansons were constantly employed on convoy protection along Bass Strait. They patrolled a beat, leaving their base at Laverton, refuelling at Bairnsdale and continuing on to Mallacoota, then stopping off again for fuel at Bairnsdale on the return. This vital, if monotonous operation was not without loss. Two of their Ansons and crews were lost during 1943. The first crashed into the sea during an anti-submarine patrol on June 21, 1943, and the other crashed at Gipsy Point on October 4, 1943.

On February 19, 1944, a lookout tower for the Volunteer Air Observer Corps was officially opened on the Esplanade at Lakes Entrance by a party of high ranking airmen, including the C.O.'s of both Bairnsdale and East Sale. Patterned on the Royal Observer Corps in England, the V.A.O.C.'s primary task was to give pre-warning of the approach of enemy aircraft coming in from the sea. They also were expected to track and assist any friendly aircraft requiring their help. The V.A.O.C. had ably performed this latter duty on more than one occasion in the Bairnsdale area over the past two years, and were to take part in a dramatic air/sea rescue only a few months after the erection of the tower.

Early in the afternoon of May 30, 1944, air observers at Lakes Entrance were alerted to keep a sharp lookout for an Anson returning from a seaward patrol but there was no sign of the plane. Later it was reported to have crashed in the sea at a position about 18 miles west of Lakes Entrance and approximately 12 miles from the shore.

It was a bitterly cold afternoon with a strong westerly wind and a choppy sea making navigation over the bar at Lakes Entrance very hazardous. Due to the weather, no boat had ventured out for the previous 12 days, but within thirty minutes of the alarm being raised, a rescue boat put out to sea manned by Mr Jack Gray with his two young sons, together with Vic. Carstairs and Harold Broome, all experienced boatmen.

After five hours sailing into the stiff headwind with very bad visibility, they reached the position of the crash, then being marked by parachute flares and surface flares dropped by search aircraft. It took a further thirty anxious minutes of searching before the cries of the four airmen, floating in their rubber dinghy, were heard over the roar of the wind and sea.

By nine o'clock that night, the aircrew and their dinghy were safely aboard the 'Lily G' and heading back to Lakes Entrance where they arrived about midnight, suffering from cold and exposure, but otherwise unhurt from the crash.

Three months later, at an Investiture at Government House, Melbourne, John, Norman and Graeme Gray and Harold Broome received the Bronze Medal of the Royal Humane Society. Victor Carstairs, who was a prior recipient of this award for work done during the 1934 floods in East Gippsland, received a bar to the medal.

Late in 1944, Wing Commander Kingwell was promoted to Group Captain and appointed commander of No. 82 (Heavy Bomber) Wing, operating B24 Liberators over the Netherlands East Indies. His place as Commanding Officer G.R.S. Bairnsdale was taken by Group Captain R.F.M. Green who had just completed an extensive tour of operations with No. 11 (Catalina) Squadron. Green, in turn, became C.O. of the R.A.A.F. Station at Pearce, W.A., and Wing Commander R.N. Dalkin, D.F.C., took up the duties of C.O. of G.R.S. He had previously flown in operations with Nos. 2, 7 and 13 Squadrons and had been Chief Instructor at G.R.S.

One of a pair of Beauforts, engaged in low-level formation practice in the area, crashed near Forge Creek early in the afternoon of September 16, 1944. The crew of two perished.

Another Anson of No. 67 Squadron force-landed in the sea off Mallacoota in November, 1944, but the crew were rescued. During 1944 enemy submarine activity off the Australian coast had tapered off. Anti-submarine patrols had been reduced until information was received towards the end of the year that a German U-Boat offensive could be expected. One U-Boat, the U.862, con ducted a series of attacks beginning off South Australia, and then continued to harass shipping along the east coast.

To meet this new threat during December, 1944 and January, 1945, Ansons of G.R.S. carried anti-submarine bombs on their exercise. They joined other aircraft from No. 67 Squadron and No. 1 O.T.U. in an effort to find and destroy the U-Boat. An ex-staff-pilot of G.R.S. recalls that one such exercise involved long hours of flying over the sea between Bairnsdale and Sydney with a full military load of fuel and anti- submarine bombs. After leaving Bairnsdale, they would turn out to sea from Lakes Entrance and fly over the ocean to cross the coast again at Twofold Bay, Eden. From this checkpoint they proceeded out to sea in a wide sweep, linking up with the coast at Moruya where the aircraft landed to refuel. From Moruya they searched the sea once again before landing at Mascot. Here the bombs were defused and removed from the aircraft by an armourer who accompanied the flight. This was necessary to allow Ansons to carry out a photographic reconnaissance of the docks and harbour installations at Sydney as no bombs were permitted to be carried by aircraft crossing the city and harbour in case of accident. Back at Mascot, the Anson would once again be bombed up for the return trip covering the reciprocal route to Bairnsdale. This, and shorter anti-submarine exercises, continued for some time until the emergency was over.

The most serious crash at Bairnsdale occurred late in the afternoon of December 6, 1944, when Beaufort A9-432 crashed while approaching to land and exploded up on impact, killing all seven occupants. This was a fully operational aircraft of No. 100 Squadron on a travel flight to Laverton. It had flown almost two thousand miles from its base at Higgins Field at the tip of Cape York and it carried the usual crew of four plus three ground staff men going on leave. This was the highest number of casualties in a single aircraft accident recorded at Bairnsdale during the war.

SIXTEEN OF OUR AIRCRAFT ARE MISSING ...

Similar words to the above were familiar to listeners of radio newscasts during the war years. The night of February 24, 1945, will be remembered by many, then stationed at G.R.S., as the night that sixteen Ansons "went missing".

On that night, a fine late summer evening, the Met. section had forecast clear conditions to continue while the night's flying programme was in progress. Within only about thirty minutes after the sixteen aircraft had taken off, a dense fog settled over Gippsland and sixteen aircraft were hopelessly lost. One pilot managed to find a hole in the fog and landed on the airfield at West Sale while the remaining crews continued to search for a landing ground until finally their fuel began to run out. Inevitably aircraft proceeded to crash-land on any available field or roadway. Considering the conditions, it was remarkable that out of about sixty airmen involved, only three were killed.

One unfortunate crew were actually on final aproach to the north/south runway at Bairnsdale when their Anson's wing collided with a tree just outside the boundary. Their aircraft crash-landed, slewed around in the opposite direction and slammed into a heavy stump, killing the pilot outright and fatally injuring the observer. Another observer, evacuating an Anson near Rosedale, left it too late to jump and his parachute did not fully open. An Anson landed on a roadway in the vicinity of Morwell, tearing off its wings on the trees bordering the road. The crew miraculously escaped injury.

Sometimes even a disaster like this contains an element of comedy. One pilot, desperately trying to keep his aircraft in the air long enough to allow his crew to jump safely, saw his observer prepare to jump, then change his mind and make his way back to find his tie, which he had discarded while working on his maps. After some rather heated instructions from the pilot, the observer hurriedly jumped. The pilot in attempting to follow, found that when he left the controls and ran down to the rear door to bail out, the Anson reared up in a steep climb preventing him from leaving. He struggled back to the controls to trim the plane into a nose down attitude and make another attempt to jump whereupon the Aggie went into a steep dive again preventing his departure. Finally, in despair, he decided that he was too low anyway and rode the crash-reluctant Anson down to a perfect landing in a South Gippsland field. This, he found later, to be the only suitable spot for many miles.

Anson N1336 was "converted to components" after running out of fuel and being forced down in Lake King, four miles east of Paynesville, on March 22, 1945. The crew were unhurt.

Many at G.R.S. were saddened by a tragedy which occurred in April, 1945.

An Anson had been detailed to fly to Lowood, Queensland, on a long reconnaissance exercise on April 7, 1945. One of the members of the W.A.A.A.F., a signals clerk, was given permission to travel with the aircraft as her husband, of only a few short months, was stationed at Lowood and due to go out on operations at short notice. While flying up the east coast, the Anson apparently ran into a violent tropical storm over northern N.S.W., was blown off course, and crashed into the sea with the loss of all the personnel aboard. Wreckage from the plane was washed up on the beach near Coffs Harbour about a fortnight later. The W.A.A.A.F. was a local Bairnsdale girl and very popular on the Station.

In a similar incident to the February crashes, Anson W1991 became lost in a sea fog on the night of June 18, 1945. After flying around for hours vainly looking for a landing field, it was ditched into two feet of water on the southern tip of Snake Island in Corner Inlet. Fears for the safety of the crew were held as the aircraft had vanished without trace, for in their endeavours to keep aloft as long as possible, the crew had jettisoned everything possible to lighten the plane. They had unfortunately included the 'Gibson Girl', the emergency transmitter, which was used in the case of a force-landing at sea. It was well into the following day before they were spotted and rescued, tired, wet and cold, but otherwise unhurt.

Others not so fortunate were the crew of Beaufort A9-143 from East Sale who were killed instantly when their aircraft dived vertically into the ground and exploded about 4 miles from Paynesville on July 30, 1945.

The early months of 1945 saw an influx of veteran crews returning from operations in Europe. As the war in that theatre was drawing to a close, Australian aircrews were coming back for rest and retraining in preparation for the final assault on Japan. There was also a surplus of aircrews held in Australia. Aircrew schools were keeping courses to a minimum and another group of personnel coming to G.R.S. were semi-trained pilots who had completed Elementary Flying Training Schools but as yet had not concluded their final training. Rather like the veterans at the other end of the scale, they were also being held in readiness for the final phase of the war.

To provide these trainees with continued air experience, several Tiger Moths were permitted to be used by them. One of the Tigers, A17-65, met an untimely end on April 23, 1945, when a pair of trainees entered a loop over Ocean Grange but finished it below sea level some distance off shore. They were lucky to escape with just bruises and a ducking.

One of the highlights in the social life of the station was the production of 'M. & V.', a 'Musical & Variety' revue produced by a very popular W.A.A.A.F. transport driver, Cpl. Rita Azzi. This show displayed the musical talents of many station personnel. It played to packed houses in the Station Cinema before repeating the performance in the Prince Regent Theatre in Bairnsdale on the evening of July 30, 1945.

Two weeks later, the war was over.

THE WAR ENDS

When the first news came through of the Japanese capitulation, V.P. Day, August 15, 1945 was celebrated from early in the morning by a tremendous barrage, over the airfield, of Verey Lights and similar pyrotechnics fired by weapons 'borrowed' from the Aircraft and Armoury Section. It was one of the most dangerous moments of the war for many, with spent, yet still red-hot, projectiles hissing down in torrents. Celebrations of a more liquid nature continued most of the day and night at the Officers' and Sergeants' Messes, and authority for once turned a blind eye to the normally prohibited consumption of alcohol in the Airmen's Mess. Much of the beer was donated by the Sergeants' and Officers' Mess committees.

With the end of the war, flying tapered off. Demobilisation began and the Station's maintenance staff, as well as aircrews, became rapidly depleted. Towards the end of 1945, flying was almost completely curtailed. Only a few travel flights were made. With the onset of another hot summer, bushfire patrols were flown to comply with the requests of the Forests Commission; sometimes an experienced forester was aboard to assist in the spotting of any possible outbreaks. It was on one of these bushfire patrols that the last aircraft accident from R.A.A.F. Bairnsdale was recorded on January 11, 1946. An Anson, MG655, supposedly spotting fires, managed to become ensnared in yards of telephone wires in the vicinity of the Marlo Hotel and ditched in the water immediately in front. The crew were unhurt and vastly amused to find that the forester aboard, who had declared in a panic that he could not swim, proved that a man can learn to swim in a hurry. He was the first to reach the shore!

At the end of December, 1945, the progressive total of pupils who had passed out of the school since its inception numbered 2,753 personnel. Most of them had been trained at Bairnsdale. The last C.O. of G.R.S. at Bairnsdale was Wing Commander A. S. Hayes, who had previously served with No. 2 (Mitchell) Squadron, Darwin.

General Reconnaissance School, Bairnsdale, was renamed the School of Air Navigation on March 13, 1946, and completed movement to East Sale on the following day.

The last R.A.A.F. unit to occupy the Bairnsdale Airfield was Care and Maintenance Unit, Bairnsdale, formed on March 1, 1946, under the command of H.Q. No. 4 Maintenance Group. Their brief was to conduct an inventory of the equipment stored on the base and safeguard it for future disposal.

Care and Maintenance Unit, Bairnsdale, disbanded on March 7, 1947, and ceased to exist as a separate Air Force unit.

With the rapid return to civilian life of thousands of airmen and airwomen, the huts and hangars of R.A.A.F. Bairnsdale (like scores of other wartime airfields) became a grey ghost town; guarded in solitude by rows of old warplanes, silent, now that they too were surplus to requirements. For many months, the Ansons and Oxfords stood there, their fabric slowly rotting in the summer sun, or lashed by the rain and hail of winter. With their tyres sagging, their metal scored by the elements and rusting in the open, they were sold as they stood, or were cannibalised for the few parts that were suitable for civilian use. Many a complete Anson found its way into someone's backyard as a child's play house, or fowl-house, for around £15 ($30). The few Tiger Moths on strength were better cared for as they stood in the shelter of the giant steel Bellman Hangars, until, they too, were disposed of at the princely sum of £75 ($150), still airworthy, but subject to a civil certificate of airworthiness. Some of this batch may still be flying. If sold to-day, they would realise prices astronomically higher than in 1947.

In August, 1946, a gigantic Auction Sale, conducted on behalf of the Commonwealth Disposals Commission, by King & Heath in conjunction with Mcarthur & Co., attracted large crowds when 1,500 lots of furniture, office equipment, tools and electrical appliances went under the hammer. A further auction in February, 1947, by Holloway Bros. for the C.D.C., cleared an additional 900 lots comprising heavy machinery, workshop tools, crockery, linen, fire fighting appliances and office furniture.

The remnants of the once proud Ansons and Oxfords were sold by the Disposals Commission in November, 1947. Among other items, the list comprised twenty-seven fuselages complete with motors, priced from only £5.10.0 each ($11), forty-four metal propellers at £1.5.0 each ($2.50), and two wooden propellers at just 10 shillings each ($1). The landing wheels could be purchased as extras with the fuselages, if required.

POSTWAR AVIATION

The expected 'Golden Age' of postwar air transportation took some years to dawn. By October, 1947, Bairnsdale Shire Council, on behalf of regional organisations, began talking of an approach to the Minister for Air to authorise an airline service to Bairnsdale. Almost three years passed before this began operating. A DC3 of Australian National Airways left Bairnsdale for Melbourne on March 27, 1950. This flight carried various dignatories and mail. The inaugural scheduled passenger flight on the Gippsland route departed Bairnsdale on April 5, 1950.

Over the past three decades, Bairnsdale has seen the rise and fall of many feeder services, some lasting years, others operating for a very short time.

After A.N.A. withdrew from the route, the passenger service lapsed for a while until Southern Airlines commenced operations in 1955. They introduced a sleek D.H. Dove on to the route, later replacing it with a larger four-engined D.H. Heron. By the late fifties they began to experience difficulties and discontinued the service in 1958.

It was not until August, 1962, that passenger flights were resumed when the Victorian Air Coach Service, a subsidiary of Ansett-A.N.A. began a daily run with a 24-seater DC3, complete with hostess. These flights lasted until February, 1966, when Ansett withdrew from the route.

Commuter Airlines commenced flying into Bairnsdale using Piaggio P166 aircraft in November, 1967. Like their predecessors, they too, found the route unprofitable and terminated the service after a year or two.

The seventies saw a succession of airlines tackle the Melbourne to Bairnsdale air route. The venerable DC3 reappeared, wearing the livery of Jetair. This continued for approximately two years. Others followed. Air Cab, owned by Executive Airlines, flew a twelve-seat Aero Commander on the route for a short time. The next to take up the challenge was East Victoria Airlines using a Cessna 402, followed by South Coast also flying a Cessna 402. Established in 1979, the current operator, North West Airlines, provides a fast, daily service, also with Cessna 402 aircraft.

Private flyers began to use the airfield shortly after the war, once fuel again became freely available. War surplus aeroplanes helped to establish many flying clubs. The Latrobe Valley Aero Club provided flying instruction on Tiger Moths, to a local flight of air enthusiasts, formed in 1953. One of the first Charter Operators was Kenair who arrived on the scene in May, 1963, and continued flying a six passenger Cessna 205 until other business commitments forced the closure of the venture in July, 1972. Local interest in flying training was revived early in the sixties when the Latrobe Valley Aero Club made week-end instruction available to another batch of trainees, this time using a Victa Airtourer and a Tripacer.

A commercial operation, the Bairnsdale Flying School, commenced pilot training and air charter work in the mid-sixties. They have continued operations in all fields of general aviation until the present. Alpine Aviation have become well established on the airfield, providing crop-dusting and aerial spraying services to district farmers and graziers.

By the early sixties, the search for oil in Bass Strait intensified and Bairnsdale was home to seven helicopters. Originally belonging to Ansett, Airfast took them over before moving to the new helicopter base at Longford. In more recent times, another helicopter company, Okanagan, is servicing the wants of Hudbay Oil in another bid for offshore oil.

Nowadays the airfield is shared by the charter operator, one or two agricultural aircraft, the oil search helicopters and third-level airline. A few small private aeroplanes seek shelter within the cavernous concrete hangars. Their cumulative monthly flying hours could not compare favourably with even one day's flying when the airfield was in its feverish infancy during the hectic days of 1942.

Like all old wartime airfields, much of the non-operational areas are now fenced off and sheep and cattle contentedly graze on close-cropped grass where once scores of service aircraft stood, lined up in regimental rows. The long, bitumened runways, that once reverberated to the incessant roar of Beauforts, Hudsons and Ansons, now infrequently awaken to the gentler drone of Cessnas, Pipers and other modern small aircraft.

Where once rows of flare pots were carefully laid out each night to line the duty runway and so guide home tired trainee crews, a modern system of runway lighting has been installed.

The concrete 'Castle', wartime nerve centre of the Station, still stands central to the airfield. Ironically, it now lives up fully to its irreverent wartime nickname, as having served for many years as a cattle shelter, the 'B.S.' is liberally coated overall.

It is the hangars that evoke the strongest memories. With great, solid concrete walls, designed to withstand almost anything except a direct hit, they were built at frantic speed during the darkest days of the Pacific war. These rare examples of the architectural school of 1942, have withstood the elements of the past four decades. They still breathe the character of a typical wartime airfield into an otherwise undistinguished landscape.

The camp area has suffered most. Where row upon row of long drab-coloured huts were home to thousands of airmen and airwomen, only a few shattered concrete bases lie scattered amid the encroaching bracken and young wattle. Here stood the canteen, the cinema, the airmen's and sergeants' messes and dozens of other buildings, all overlooked by the officers' mess crowning the low ridge behind the oval "Killingsworth Centre." The station sickquarters alone has survived the peace and is now home for the aerodrome groundsman. The other hutments were sold off after the war, many becoming farm sheds, bungalows, garages, storerooms or temporary class-rooms at district schools. Several of the buildings went to the Woomera rocket range.

The personnel who had occupied those huts, sometimes remained for long periods as staff, or on instructional duties. Some spent only a few short weeks during their final phase of training — learning to fly and fight as a crew — or to undertake specialised courses before departing for the war zones.

Many, tragically, lost their lives in aircraft accidents, or disappeared forever into the deep, dark waters of Bass Strait.

It is to all men and women who served at R.A.A.F. Bairnsdale, that this book is dedicated.

A striking photograph of a Lockheed Hudson taking off. About 250 Hudsons served with the R.A.A.F. during the war on a multitude of duties. The first unit, based at Bairnsdale, No. 7 Squadron, had six Hudsons initially and No. 1 O.T.U. operated 15 of these aircraft. Several were lost at sea or in crashes in the Bairnsdale district.

THANKS!

The author has talked about aviation in general and R.A.A.F. history in particular, with so many people over the past several decades, that to make a full list of acknowledgements would not be possible. He would like, however, to especially thank the following organisations and individuals who provided so much of the information on which this book is based, or otherwise encouraged the author to carry on with research:-

The Australian War Memorial; "Bairnsdale Advertiser"; Bairnsdale Shire Council; The "Gap" Magazine; R.A.A.F. Historical Section; 7 Squadron Association; W. L. Allen, ex-1 OTU; N. Broughton, ex-1 OTU; Mrs June Bishop, ex-GRS; Norm Christie; Ray Dorrington; Alan Fraser, ex-GRS; Stan Harris, ex-1 OTU; Mrs Margaret Hunt, ex-1 OTU; Mr and Mrs Bob Hussey; Vic. Jones, ex-7 Sqdn.; Bob Kenworthy; Mrs M. Lillicrapp, ex-1 OTU; Bill Lillicrapp, ex-1 OTU; Col. McPherson, ex-GRS; Jack Newton, ex-GRS; Chris Rawlings; Bert Rodd, ex-1 OTU Special Transport Flt.; Cecil Stoneham, dec'd; Ron Yeates; . . . . and my wife, Lorna, who has put up with reams of loose notes scattered throughout the home and turned many an ungrammatical phrase into English in the final manuscript.

Ansons of No. 67 Squadron, shown at Laverton, used Bairnsdale as an advanced base during convoy patrols from Laverton to Mallacoota. The Anson on the left, R9935, served with G.R.S. from September, 1943, until May, 1944.

Line-up of Avro Ansons at G.R.S. Bairnsdale late in 1945. Many Ansons were sold by the Commonwealth Disposals Commission after the war, for as little as £15. Incomplete Ansons were finally cleared for £5.10.0 in 1947.

The Unit Cinema was the centre of entertainment on the Station, where films were shown each night of the week. Camp Revues and airmen's dances were also held periodically.

The Sergeants' Mess catered for over 300 sergeants and warrant officers. The sergeants' living quarters are behind the Mess and the Guardroom is on the extreme right.

The Farewell Party held in the Sergeants' Mess in December, 1945. Several Warrant Officers were Regular Airmen and continued to serve at East Sale. The majority of sergeants in the photograph were demobbed by early 1946.

..... AND NOW:— (A) The only surviving building in the Barracks Area dating from the war, is part of the Station Sick Quarters, now occupied by the Aerodrome Groundsman and his family.

(B) The Bomb Dump still stands practically unchanged in a quiet tree-lined lane, just outside the south-eastern corner of the airfield.

..... AND NOW: — (C) The solid concrete Ammunition Dump on the Eastern perimeter of the airfield is pockmarked by bullet strikes, not from the war, but from its use as a prime target of postwar shooters.

(D) The Entrance to the Emergency Operations Room, situated on private land about 2 miles from the airfield. This was buried deep under-ground and designed to withstand the heaviest air raid and still remain operational. Fortunately it was never put to the test.